omar FARSSI

Optimization of irrigation techniques

omar FARSSI

Optimization of irrigation techniques

Impact of various irrigation strategies on the agro-physiological parameters of a young olive plantation.

ScienciaScripts

Imprint

Any brand names and product names mentioned in this book are subject to trademark, brand or patent protection and are trademarks or registered trademarks of their respective holders. The use of brand names, product names, common names, trade names, product descriptions etc. even without a particular marking in this work is in no way to be construed to mean that such names may be regarded as unrestricted in respect of trademark and brand protection legislation and could thus be used by anyone.

Cover image: www.ingimage.com

This book is a translation from the original published under ISBN 978-620-6-72704-0.

Publisher:
Sciencia Scripts
is a trademark of
Dodo Books Indian Ocean Ltd. and OmniScriptum S.R.L publishing group

120 High Road, East Finchley, London, N2 9ED, United Kingdom
Str. Armeneasca 28/1, office 1, Chisinau MD-2012, Republic of Moldova, Europe
Printed at: see last page
ISBN: 978-620-3-74494-1

CONTENTS

SUMMARY

This study focused on the effect of the irrigation regime on certain agro-physiological and biochemical parameters in a young olive tree plantation. The trial was set up at the Domaine Expérimental de Saada of INRA Marrakech. Three irrigation regimes were considered: T1= (100% ETc), T2= (70% ETc) and T3= Farmer's Practice (FP). The irrigation system used for the first two regimes was drip irrigation, while the PA treatment used gravity-fed irrigation. ETc was calculated on the basis of meteorological data from the experimental station. The results obtained showed that the PA irrigation regime affected the agro-physiological and biochemical parameters measured compared with the 100% ETc irrigation regime. Significant reductions in vigour parameters were noted under this irrigation regime. Water saturation deficit and membrane permeability were found to increase under gravity irrigation indicating the detrimental effect of this regime. Stomatal conductance and leaf chlorophyll content were negatively affected. As a result, the high quantities of irrigation water provided in this treatment (PA) are not used efficiently, which has induced water stress in young olive trees. As for the water stress induced by the deficient irrigation in treatment T2 (70% x ETC), it did not induce significant negative effects on the agro-physiological parameters analysed compared with treatment T1 (100% ETc). Hence the importance of adopting this irrigation regime and saving 30% of water used compared with T1 (100% ETc) and 80% of water used compared with T3 (PA).

Key words: Young olive trees, deficit irrigation, drip irrigation, gravity irrigation, agro-physiological parameters, water deficit, evapotranspiration.

INTRODUCTION

In many parts of the world, there is growing interest in the olive tree and its products (Fernandez and Moreno, 1999). Indeed, the olive tree is one of the most interesting species for cultivation in arid and semi-arid zones. This is due to its remarkable adaptation to drought, which allows it to develop and produce under rainfed conditions in areas with average rainfall, even less than 500 mm per year, and where the dry season can last five or six months. The agronomic interest of the olive is reinforced by the fact that it shows a remarkable response to any improvement in growing conditions. The increase in irrigated olive-growing area will therefore give rise to a conflict of interest in the use of water in relation to other crops and other uses, due to the scarcity of water resources. Like many countries around the world, Morocco is facing the problem of developing and managing water resources sustainably. In agriculture, and more specifically in olive growing, it is necessary to optimise irrigation by estimating water requirements accurately. To achieve this objective, it is necessary to control the water balance of the soil and certain eco-physiological parameters related to the olive tree. The aim is to improve the economical management of water inputs and increase the efficiency of irrigation water use. Deficit irrigation is one of the remedies adopted. It is based on improving growth while reducing evapotranspiration losses (Centritto et al., 2000; Costa et al., 2007). It has been applied to different species in different climates, and helps to improve water conservation and use efficiency.It is within this framework that this study was carried out, with the aim of studying the response of a young olive tree plantation to deficit irrigation compared with the conventional irrigation technique. To do this, we evaluated the effect of three water regimes on certain agro-physiological parameters of young olive trees. The irrigation treatments studied were:

- Gravity irrigation used by farmers

- Localised irrigation corresponding to 100% ETc

- Localised irrigation deficit corresponding to 70% ETc

CHAPTER I
STUDY BIBLIOGRAPHY

I. General information on the olive tree

I.1. Botanical history and description

The olive tree originated in Asia Minor, where it is abundant and forms a veritable forest. It seems to have spread from Syria to Greece (De Candolle, 1883). Although there are other hypotheses in lower Egypt, in Nubia. This is why Caruso considers it to be indigenous to the Mediterranean basin. From the 6th millennium onwards, its cultivation gradually spread throughout the Mediterranean basin (Camps, 1974). Olive cultivation is also very old in Morocco. The Moroccan picholine that dominates Moroccan olive groves appears to be a hybrid between an eastern and a western maternal line (Boulouha et al., 2006). The trunk of the tree is grey-green until its tenth year and varies in height from 4 to 8 metres depending on the variety (Barranco et al., 1995). It is generally circular but, with age, takes on a twisted appearance, characteristic of the tree (Loussert et Brousse, 1978). The tree is distinguished from other fruit species by its longevity. The development of the root system depends on the physico-chemical characteristics of the soil: olive tree roots are able to adapt to the depth of the soil, depending on its structure and texture (Loussert et Brousse, 1978). The leaves of the olive tree are simple and opposite, with a shiny, leathery upper surface that is dark green and shiny due to a waxy layer that limits evapotranspiration, and a silvery underside (Amouretti and Comet, 2000). The stomata are sunken into crypts, reducing gas exchange (Loussert et Brousse, 1978). The flowers are long, flexuous clusters with four to six secondary branches (Loussert et Brousse, 1978; Boulouha et al., 2006). The

5

same inflorescence has hermaphrodite flowers with an ovary surmounted by a short style and a broad stigma and imperfect flowers without pistils (Boulouha et al., 2006). The fruit is an elliptical, globular drupe measuring 1 to 4 cm in length and 0.6 to 2 cm in diameter and is green in colour, becoming black to purplish-black when ripe. The epicarp is always attached to a pulp, the mesocarp, which is fleshy and rich in fat (Barranco et al., 1995 ; Loussert et Brousse, 1978).

I.2. Classification

The olive tree belongs to the Oleaceae family, which comprises 29 genera. The genus Olea europea L. contains around 30 species. It belongs to the spermaphytes phylum, angiosperms subphylum, ligustrales order, Oleannae tribe (Loussert et al., 1978).Within the genus O europea, there are two subspecies:

▶ Olea europea oleaster or silvestris L. or wild olive, well adapted to the rigours of the Mediterranean climate (Boulouha, 1991).

▶ Olea europea sativa: The cultivated olive tree characterised by its longevity and durability (Boulouha, 1991).

I.3. Vegetative and productive cycle of the olive tree

During its annual development cycle, the olive tree goes through the following phases (Farouk, 2011):

❖ January, February: induction, initiation and floral differentiation.

❖ During March: growth and development of inflorescences in the leaf axils of the previous year's shoots.

❖ Mid-April: full flowering.

❖ Late April-early May: Fertilisation and fruit set.

❖ June: Fruit begins to develop and grow.

❖ September: veraison.

❖ October: Ripening of the fruit and enrichment with oil.

❖ Mid-November to January: fruit harvest.

The most intense period of the annual cycle is from March to June. During this phase, the tree's need for water and nutrients is at its highest.

I.4. Soil and climate requirements of the crop

I.4.1. Climatic requirements

The olive tree can withstand temperatures of -8 to -10°C in winter and 40°C in summer, but these extreme conditions have a negative impact on the tree's flowering and fruiting. Temperatures play a decisive role in the flowering of this species, which forms its flower buds at the end of winter, as the olive tree needs to be exposed to the sun for around ten weeks in order to form its flower buds. Light is also a limiting factor, as insufficient illumination affects fruiting and encourages the development of pathogens (Boulouha et al., 2006, M.A.D.E.R., 2003).

I.4.2. Soil requirements

The olive tree is a hardy tree that grows in different types o f soil, but prefers deep, permeable, rich and well-balanced soils (Boulouha et al., 2006; M.A.D.E.R., 2003). Winds are harmful because of their mechanical action, which can b e more noticeable during flowering and when the fruit is close to ripening (C.O.I. l'olivier l'huile l'olive).In addition, the tree tolerates a good pH margin, provided that the pH is not below 6.5 to avoid releasing magnesium or aluminium ions that are toxic to the plant. In addition, at high pH (>8.3), phosphorus and iron tend to become insoluble (C.O.I., 2007).

I.5. Socio-economic importance of the olive tree in Morocco

The olive sector contributes 5% to the national agricultural GDP. Covering an area of over one million hectares, the country's olive farms produce around 1,500,000 tonnes of olives. The country also produces 160,000 tonnes of olive oil and 90,000 tonnes of table olives. In terms of exports, 17,000 tonnes of olive oil and 64,000 tonnes of table olives are sold on international markets (MAPM, 2013). Olive trees are grown on 784,000 hectares of land, covering all of the country except the Atlantic coastal strip (MAPM, 2006).The olive-growing sector is a major agricultural activity, generating more than 15 million working days a year. This sector makes a significant contribution to the income of a large fringe of poor farmers, while also playing a key role in feeding rural populations (MADRPM, 2008).

II. Olive tree irrigation

II.1. Irrigation methods

Irrigation is the operation of artificially supplying water to cultivated plants to increase production and enable them to develop normally in the event of a water deficit caused by rainfall, excessive drainage or a drop in the water table, particularly in arid areas (Loussert and Ferrak, 2011).The irrigation methods used in olive growing include gravity irrigation and localised irrigation.

II.1.1.Gravity or trickle irrigation

It consists of distributing water over the cultivated plot by run-off onto the soil in the furrows or by controlled submersion. It remains the most widespread method of irrigation in the world. In Morocco, it is estimated that over 93% of the surface area of large hydraulic systems are irrigated using a traditional technique called "Robta".(Azouggagh, 2001).This type of irrigation is not very efficient because it wastes a lot of water through evaporation and limits the frequency or duration of the irrigation period because of the high doses required. Nevertheless, these methods offer the advantage of low investment in equipment and are not very costly (Azouggagh, 2001). This may explain why they are so widely used in olive growing.

II.1.2.Localised irrigation (drip irrigation)

Drip irrigation is the most widely used method of localised irrigation. It has enjoyed considerable growth since the end of the 20th century[ème] thanks to the water savings it allows (30% compared with conventional sprinkler irrigation) (Lamine, 1993).Given its advantages, localised irrigation is a high-performance, sophisticated technique whose application is generally indicated for intensive orchards in arid and semi-arid zones (Filali, 2010).This method of irrigation can be described as the frequent application of water in very low doses. It consists of transporting filtered, treated water, which may be enriched with fertilising elements, through a system of pipes and tubes, often made of plastic (PVC and PE), from the source to the vicinity of the plant. This is remarkably effective because the soil solution near the roots is more concentrated, which means a higher yield because transpiration efficiency is high. This water is applied to the root zone using a device called a dripper, specially designed to discharge water at very low flow rates of the order of a few litres per hour at low pressures (around one bar). In this method of irrigation, the volume of water stored in the soil is not taken into account (Filali, 2010).

The advantages of this technique include

❖ Great water savings: young drip-irrigated orchards use no more than half of what is normally consumed when they are irrigated with a conventional sprinkler system. aspersion or gravity. The technique owes this saving to the efficiency of the network, the absence of losses through percolation and run-off, and the reduction of losses through evaporation and weeds.

❖ Saves energy and money thanks to reduced labour requirements.

❖ Extremely easy to manage: the soil is only partially wetted, which makes it possible to easy access to the field for other cultivation techniques. Fertilisers can thus be injected directly into the root zone with greater precision and efficiency (Filali, 2010).

There has been a certain amount of interest in converting from gravity-fed irrigation to localised irrigation in Morocco. In 2006, farmers equipped around 141,000 ha with localised irrigation, i.e. 9.7% of the total developed area (Bekkar et al., 2007).

II.2. Water requirements of olive trees

The olive tree is a species that is relatively resistant to water stress. This plant is characterised by a number of anatomical adaptations and physiological mechanisms that enable it to preserve its vital functions, even under very severe conditions. The study of determining the water requirement of the olive tree on the basis of the ETP value in olive-growing regions. This value can be estimated using various formulae. In the course of several experiments, it appeared that olive growing could consume almost 100% of the ETP (Vernet and Mousset, 1963). These authors also arrived at very high ETM values during the winter (80%). The water requirements of olive trees in a semi-arid environment have been estimated at 65% of potential evapotranspiration (ETP) (Boulouha et al. 2006). These needs are assessed on the basis of actual evapotranspiration (AET= ETP x Kc). Potential evapotranspiration is the evaporative capacity of the atmosphere on soil with plant cover that has an abundance of water. It is directly linked to climatic conditions.The crop coefficient (Kc) must be determined experimentally, and varies from region to region (Orgaz and Fereres, 2007). In the Haouz region, the Kc is between 0.6 (between September and May) and 0.7 (between June and August) (ORMVAH, 2005).To find out exactly how much water each olive tree should receive, you need to is based on the following data:

• Evapotranspiration AND_0 Given by weather stations (taking into account temperature, radiation, wind speed and direction).

• Rainfall.

• The diameters of the tree canopy (D1 and D2) measured using a yardstick graduated.

• Planting density.

Crop evapotranspiration (ETc) can therefore be expressed by the following formula (Orgaz and Fereres, 2007):

ETc = ((ET_0 x Kc x Kr) / ER)-PE

Kc: 0.6 or 0.7 depending on the season

RE: network efficiency

PE: effective rainfall

Kr: (reduction coefficient) determined by the formula of Fereres et al. (1981, 1984, 1996) Kr= 2 (Sc /100)

Sc (area covered by olive trees) = (rc × D2 ×N) /400

D: average diameter of the foliage

N: plant density

III. Water stress and its effect on plants

III.1. General information on water stress

Nowadays, drought is considered to be a stress with a major impact on crop plants. As far back as the early 1980s, Christiansen (1982) estimated that almost half of the land available for agriculture was threatened by a shortage of good quality water. Similarly, Bray (1997) reported that 40% to 60% of agricultural land is exposed to drought. We can speak of water stress when there is an excess of water (flooding causing asphyxiation) and when there is a lack of water (William and Charles-Marie, 2003). A water deficit occurs when the quantity of water transpired by the leaves is greater than the quantity of water absorbed by the roots (Trossat, 2005).

III.2. Effect on stomatal conductance

Stomatal regulation is a crucial point, determining not only the flow of transpiration, but also photosynthesis and the flow of assimilates to the various parts of the plant. Stomatal conductance indicates the rate of leaf transpiration. During a water deficit, stomatal conductance will decrease (Kotchi, 2004).

III.3 Effect on fruit growth

Aggabio (1974) shows that although the growth curves are the same as in dry cultivation, the final size of the fruit is larger. Baldini and Pisani (1963) believe that late irrigation conditions the final size of the fruit, and is therefore recommended for table varieties.

IV. Mechanisms of tolerance to water stress in plants

IV.1. Biochemical adaptations

IV.1.1. Accumulation of soluble sugars

Carbohydrates are important sources of reserves in plants (Binzel et al., 1987). Under conditions of environmental stress, these compounds act as osmotic effectors to maintain an adequate osmotic potential and also as protective molecules for membrane structures and proteins against dehydration (Hoekstra et al., 2001; Mckersie and Leshan, 1994). In addition, a significant accumulation of soluble sugars has been reported in several species in the presence of water stress, such as soya (De Ronde et al., 2004; Muller et al., 1996).

IV.1.2. Proline accumulation

Proline accumulation is linked to osmotic adjustment during a water deficit (Ashraf and Harris, 2004).In plants under stress, Kemple and Mac Pherson (1954), Barnett and Naylor (1966) and Hubac and Chouard (1973), in Nemmar, (1983) show an increase in proline concentration and note that this accumulation is a direct consequence of the water deficit. Hubac (1967) and Le Saint (1969) show that exogenous proline applied to seedlings increases their resistance to drought.

IV.1.3. Protein synthesis

Drought conditions bring quantitative and qualitative changes to plant proteins (Seyed et al., 2012).The induction of proteins by water deficit depends on the stage of development of the organ in question and the genotype (Riccardi et al., 1998; Vincent et al., 2005).Overall, the studies already carried out show that resistance to water deficit in plants is a highly complex phenomenon, with a cascade of reactions involving a very large number of genes and gene products.

CHAPTER II
MATERIALS & METHODS

I. Presentation of the experimental site

This study was conducted in the field at the Saâda experimental estate of the Institut National de la Recherche Agronomique, located 7 km west of the city of Marrakech, in an agro-ecological zone at an altitude of approximately 468 m. The trial was conducted on clay-loam soil, with mean annual maximum and minimum temperatures of 28.7°C and 13.5°C respectively, and total annual rainfall of 226.54 mm. In addition to rainwater, the Sâada site has water from a dam (flow rate of 30 l/s) and two 150 m-deep boreholes. Irrigation is currently provided by boreholes.

II. Plant material and experimental set-up

This study focused on young olive trees of the 'Menara' variety planted in December 2010 at the Domaine Expérimental in Saada. The trial was conducted on two plots (Figure 1). The first with a localised irrigation system (drip) and the other with a gravity irrigation system.In the first plot, the system is based on complete random blocks. Each block has two irrigation doses: one dose (T1) corresponding to 100% ETc and a deficit dose (T2) corresponding to 70% ETc. On the second plot, irrigation is by gravity system with a regime (T3) corresponding to the inputs used by farmers in the region.

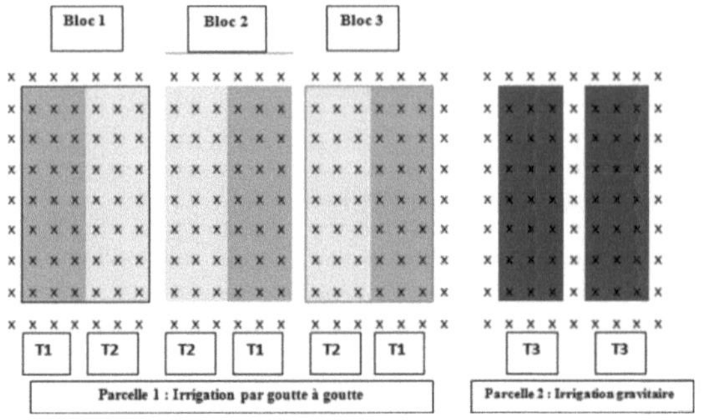

Figure 1: plan of irrigation trial on a new olive plantation

x : Tree

T1: Trees irrigated by drip (100% ETC): 190 mm T2: Trees irrigated by drip (70% ETC): 133 mm

T3: Trees irrigated by the gravity-fed system used by farmers: 675 mm

Each colour delimits the trees in each water regime studied, with the rest of the trees representing the edges.

Figure 2: Photo of the trial

III. Evaluation of the trial

III.1. Agronomic parameters

III.1.1. Vigour

Vigour was studied on all the trees (168 trees). The measurements taken were :

► Total height (Ht)
► The height of the foliage (H)
► The maximum diameter of the foliage (D1)
► Minimum canopy diameter (D2)
► The perimeter of the trunk: measured at 10cm from the ground (P)

III.1.2. Fructification

At the end of June, which corresponds to the fruiting stage, we counted the number of trees that had fruited for each water regime, so that we could determine the percentage of fruiting trees per water regime.

III.2. Physiological and biochemical parameters

III.2.1. Water saturation deficit (HSD)

Plant tissues are most often in a state of water deficit and the DSH makes it possible to assess this state (Coudret, 1981). To determine this parameter, batches of leaves were taken and their fresh weight (fw) was determined. The leaves were then soaked in distilled water for 12 hours at room temperature to reach water saturation. The leaves were then wiped with blotting paper and their saturated fresh weight (PFsat) was determined. They were then dried for 24 hours at 80°C and their dry weight (DW) was determined. The DSH is defined as follows:

DSH % = ((PFsat- PF) / (PFsat -PS))*100)

III.2.2. Membrane permeability

The effect of irrigation regime on membrane permeability was assessed by the percentage of electrolyte loss (Lutts et al., 1996). Leaves were rinsed several times with distilled water and placed in flasks containing 10 ml of distilled water. The electrical conductivities (C_1) of the solutions were then determined using a Hanna HI 8733 conductivity meter after incubation for 24 hours at 25°C with stirring at 100 rpm. A second conductivity (C_2) of the samples was determined after autoclaving at 120°C for 20 min, followed by shaking for 30 min at 25°C. The percentage electrolyte loss was expressed as the ratio of C_1 and C_2

Electrolyte loss (%) = (C /C_{12})*100)

III.2.3. Stomatal conductance

Stomatal conductance indicates the rate of leaf transpiration, which is a parameter linked to the plant's water status, as it signals the opening or closing of stomata. The effect of the irrigation regime on stomatal conductance was assessed on mature leaves using a leaf porometer. We carried out 3 repetitions per tree.

III.2.4. Determination of soluble sugars

The carbohydrates are dehydrated in sulphuric acid at high temperatures to form furfural derivatives, which combine easily with phenol to give a pinkish-salmon colour.Ethanol can be substituted for MCF for extraction using the AOAC (Association of Official Analytical Chemists) method modified by Nguyen and Paquin (1971). 0.5 g of leaves were homogenised with 5 ml of 95% ethanol. The upper phase of the filtrate was separated and the sediment was washed twice with 5 ml of 70% ethanol and the upper phase was added to the previous one. The mixture was centrifuged at 3500 g for 10 min at 4°C and the supernatant of the alcoholic extract obtained was recovered and stored in a refrigerator at 4°C overnight as for extraction with MCF (Paquin and Lechasseur, 1979). Ethanol has the advantage over MCF of allowing the determination of sugars, organic acids and other soluble compounds in the same extract.One tenth of a ml of the alcoholic extract was mixed with 3 ml of Anthrone (150 mg Anthrone, 100 ml 72% v / v sulphuric acid). The samples were placed in a boiling water bath for 10 minutes. The optical density of the samples was assessed at 625 nm using a spectrophotometer. The content of soluble sugars was determined using the standard glucose range and expressed as mg g^{-1} of leaf DW.

Preparation of the standard range :

The various concentrations are prepared from a 0.2 g/l glucose stock solution (Table 2):

Table 2: Preparation of the standard range for the determination of soluble sugars

Glucose concentration	0 µg/ml	20 µg/ml	40 µg/ml	60 µg/ml	80 µg/ml
Distilled water (ml)	10	9	8	7	6
Glucose stock solution (ml)	0	1	2	3	4
Anthrone	3ml				
Boiling at 85°C (ml) for 10 min					

III.2.5. Proline assay

Extraction is carried out using the AOAC method as for soluble sugars. Using the modified method of Singh et al (1973), an aliquot of 0.2 to 1 ml of the upper phase is pipetted off, depending on the concentration of proline. Care is taken to avoid touching the lower phase, so 5 ml of water is added, followed after stirring by 2.5 ml of ninhydrin (0.125 g in 2 ml of $H_3 PO_4$ 6M plus 3 ml of glacial acetic acid) and 2.5 ml of glacial acetic acid. After stirring, heat on a water bath at 100°C for 45 min. The mixture was cooled and 1ml of toluene was added. The proline-ninhydrin complex formed was extracted by adding 1 ml of toluene to the tubes after cooling.After stirring, we left it to stand for 30 minutes. The optical density of the upper phase was measured using a spectrophotometer at 520 nm. The concentration of the proline content was determined using a standard range made under the same conditions from different proline concentrations.

Preparation of the standard range :

From the proline stock solution (100mg/l), a 20µg/ml daughter solution was prepared by suspending 100ml of the stock solution in 400ml of distilled water. A series of dilutions of 15, 10 and 5µg/l were made from the proline daughter solution (Table 4).

Table 4: Preparation of the standard range for the proline assay.

Proline concentration	0 µg/ml	5 µg/ml	10 µg/ml	15 µg/ml	20 µg/ml
Distilled water (ml)	10	7,5	5	5	0
Proline daughter solution (ml)	0	2,5	5	7,5	10
Ninhydrine	2,5ml				
Acetic acid solution (ml)	2,5ml				
Boiling at 85°C (ml) for 45 min					

III.2.6. Protein assay

The proteins were assayed using the Bradford method (1976). This is a colorimetric assay, based on the change in absorbance, manifested by the change in colour of Coomassie Blue G250 after binding with the aromatic amino acids (tryptophan, tyrosine and phenylalanine) and the hydrophobic amino acid residues present in the protein(s).Extraction is by cold grinding 100 mg (MF) in 4 ml of 0.1 M Tris-HCl buffer at pH 7.5. The crushed material was then centrifuged at 16,000g for 15 min. The supernatant was recovered and the pellet transferred to 2 ml of buffer and centrifuged. The two supernatants were mixed and used for the determination of soluble proteins. 2ml of diluted extract was taken in tubes and 2ml of Bradford reagent was added. After shaking followed by a 4 min rest at room temperature, the density The optical density (OD) was measured at 595 nm. Protein levels were determined using a standard range established by bovine serum albumin (BSA) solutions.

Preparation of the standard range

Stock solution of SAB (**Bovine serum albumin**) (10mg/100ml) (table 3)

Table 3: Preparation of the standard range for the protein assay

SAB µg/ml	5	10	20	30	40
Stock solution	0.5	1	2	3	4
Distilled water	9.5	9	8	7	6
Reagent for Bradford	2ml				
	Rest for 4 min and optical density at 595nm				

III.2.7. Chlorophyll content

The chlorophyll was extracted using the method of Arnon (1949), by grinding 100 mg of fresh material in the presence of 2.5 ml of 80% acetone. After centrifugation (10min, 5000×g), the optical density was measured using a spectrophotometer at 663 and 645 nm. The chlorophyll content was determined using the following formula:

Chl (a + b) = 8.02 (OD at 663) + 20.20 (OD at 645) mg.ml^{-1}

IV. Statistical analysis

Statistical analyses were carried out using SPSS version 10 software, using the ANOVA test and the SNK (Student-Newman-Keuls) test to determine homogeneous groups when the effect of the treatments studied was significant.

CHAPTER III

RESULTS & DISCUSSION

I. Effect of irrigation regime on agronomic parameters

I.1. Effect on the vigour of young olive trees

I.1.1. Effect on total height and canopy height

Figure 3 shows the effect of irrigation regime on the height of young olive trees.

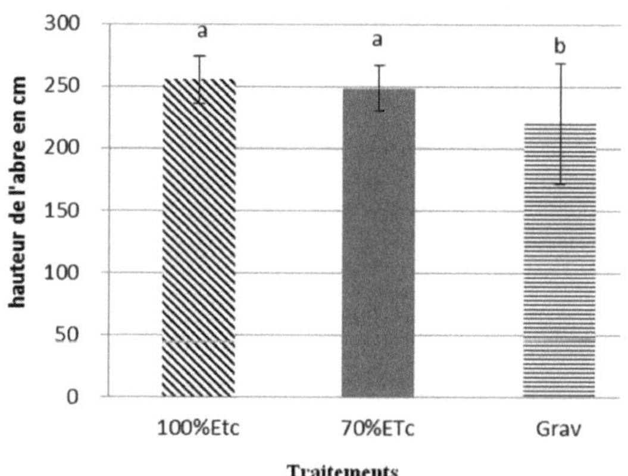

Figure 3: Effect of water regime on the height of young olive trees

The average height of the trees varied according to the irrigation regime. This variation was highly significant (P<0.001 at 2 ddl) (Table 1, Appendix). The SNK test (Appendix Table 10) classified the treatments studied into two homogeneous groups, the first group consisting of the two regimes T1 (100% ETc) and T2 (70% ETc) and the second group consisting of treatment T3 (PA). Consequently, there were no significant differences between the two treatments (T1 and T2). The latter two differed significantly from treatment T3 (PA). These

23

reductions in growth linked to the water regime are associated with a decline in cell enlargement (Bhatt and Srinivasa, 2005). Under water stress, cell elongation in plants can be inhibited by the interruption of water flow from the xylem to elongating cells (Nonami, 1998). Similarly, water stress reduces photo-assimilation and the metabolites required for cell division. As a result, mitosis is reduced, as is cell elongation and expansion, leading to reduced growth (Farooq et al., 2009).

I.1.2. Effects on average tree diameters

Figure 4 shows that the average diameter of the trees varied according to the irrigation regime. Analysis of variance showed a highly significant effect of irrigation treatments (P<0.001 at 2 dll) (Table 2).

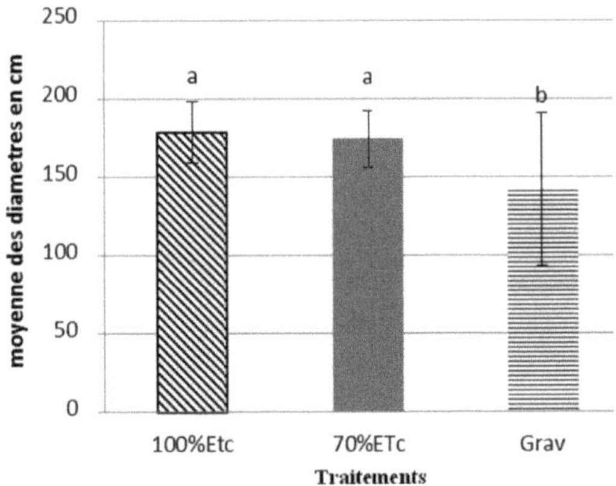

Figure 4: Effect of water regime on the diameter of young olive trees

The SNK test (Table 11) classified the treatments studied into two homogeneous groups, the first group consisting of the two regimes T1 (100% ETc) and T2 (70% ETc) and the second group consisting of the gravity treatment T3 (PA). We noted that treatment T1 (100% ETc) recorded the highest average diameter growth rate (178.5m), followed by treatment T2 (70% ETc) (174.3m) and finally treatment T3 (PA) (142m). Consequently, we did not note any significant differences between the two treatments (100% ETc and 70% ETc). The latter two treatments differed significantly from the PA treatment. This reduction in tree diameter under the influence of water stress was reported by (Fereres, 1984).

I.1.3. Effects on the perimeter of tree trunks

The perimeters of the tree trunks according to the irrigation regime are shown in Figure 5.

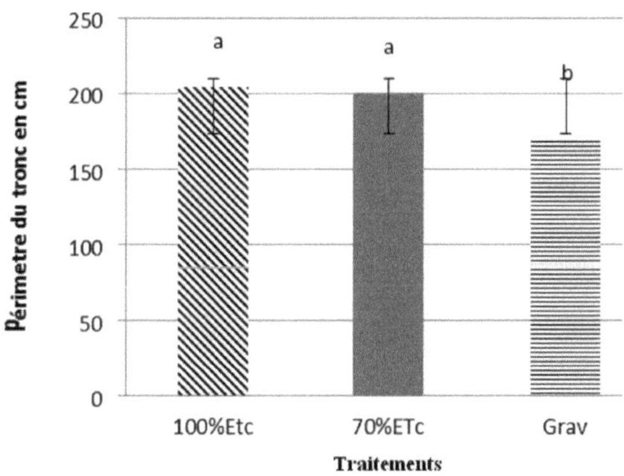

Figure 5: Effect of water regime on trunk perimeter of young olive trees

This parameter varied according to the irrigation treatments studied. The variations noted between the different treatments were found to be highly significant b y   analysis of variance (P<0.001 at 2ddl) (Table 3), and the highest values for trunk cross-sectional area were obtained by treatment T1 (100% ETc). Multiple comparison of the means using the SNK method (table 12 in the appendix) classified the treatments studied into two homogeneous groups. The two regimes 100% ETc and 70% ETc formed a single homogeneous group, while the PA regime formed the other group. Consequently, the effect of these two irrigation regimes on the growth rate of the trunk section area is identical. This reduction is in line with the literature, which mentions a significant reduction in trunk growth expansion due mainly to a negative influence of water restriction on the physiological and biochemical processes of young trees (Ennajah, 2006).

I.2. Effects on tree fruiting

Figure 6 shows the rate and extent of fruiting obtained for each treatment.

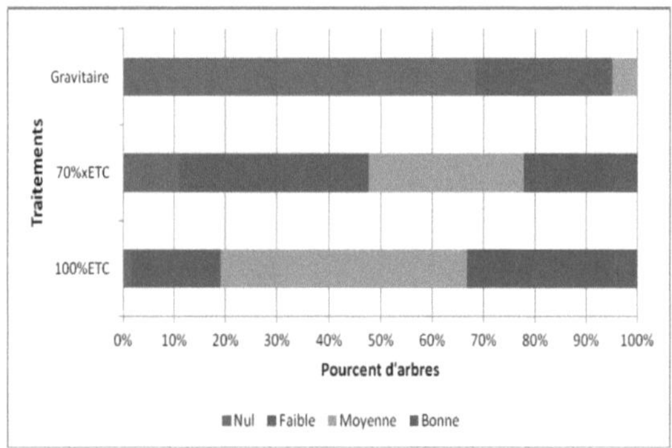

Figure 6: Rate and extent of fruiting of young olive trees obtained for each treatment.

Under the T3 (PA) regime, only 32% of the trees produced fruit, compared with 98% and 89% for the T1 (100%ETc) and T2 (70%ETc) treatments respectively. In addition, 33% and 22% of the trees produced good fruit under the T1 (100%ETc) and T2 (70%ETc) regimes respectively. In the T3 (PA) regime, fruiting was only average in 5% of trees and poor in 27% of trees. This can be explained by the fact that the T2 (70%ETc) deficit irrigation does not induce severe water stress in young olive trees compared with the T3 (PA) treatment.

II. Effects of irrigation regime on physiological parameters.

II.1. Effect on water saturation deficit (HSD)

The average DSH values of the highest leaves were obtained with the gravity irrigation regime, whereas the minimum values of this parameter are represented by the 100% ETC regime (Figure 7). These differences between the three treatments were revealed to be significant by analysis of the variance. (Table 5 appendix)

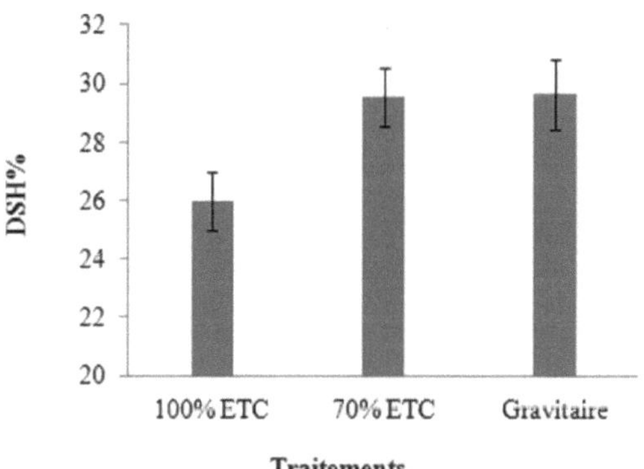

Figure 7: Effect of water regime on the water saturation deficit (HSD) of young tree leaves olive tree

The water saturation deficit values are complementary to the relative water content values. The significant difference between these two treatments and the PA treatment can be explained by the exposure of young trees to water stress. Kasraoui et al (2004) have shown that water status in olive trees is affected by the level of water restriction.

II.2 Effect on electrolyte loss

The data in figure 8 showed that the irrigation regime had a highly significant effect (P<0.001) on electrolyte loss in young olive trees. The lowest values of this parameter were recorded under 100% ETC irrigation (10.98%) and the highest values were recorded under gravity irrigation (20.55%). On the other hand, under 70% ETC irrigation, electrolyte loss was 11.64%.

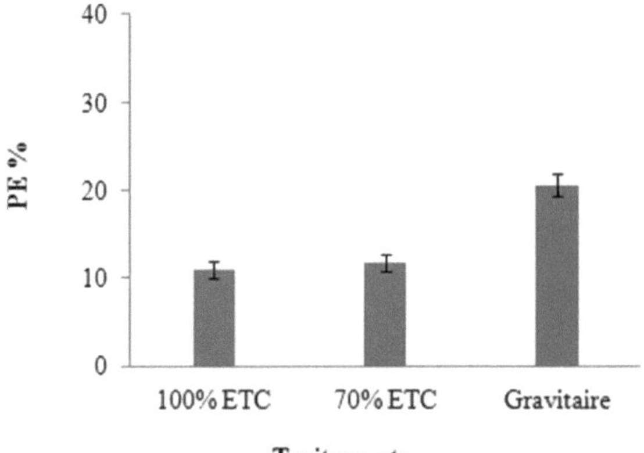

Figure 8: Effect of water regime on electrolyte loss from leaves of young olive trees

Increased electrolyte loss indicates destabilisation of cell membranes. Similar results have been reported in alfalfa (Farissi et al., 2013).

II.3. Effect on stomatal conductance

The values of stomatal conductance measured at the level of the leaves of young olive trees are shown in figure 9.

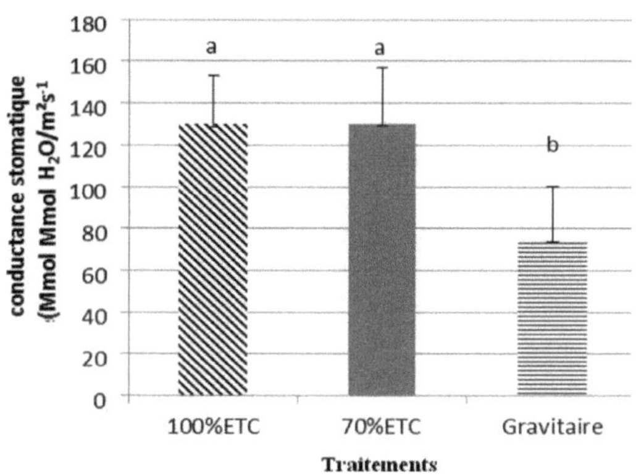

Figure 9: Effect of water regime on stomatal conductance in young olive trees

This parameter varied according to the treatments studied, and the analysis of variance showed a high significant difference between the 3 treatments (P<0.001 at 2 ddl) (table 4, appendix).Treatment T3 (PA) induced the lowest values of stomatal conductance (75 mmol of H_2 O/m².s) and the highest value of this parameter was obtained by treatment T2 (70% ETc) (130 mmol of H_2 O/m².s). The multiple comparison of averages using the SNK method (table 13 in the appendix) classified the treatments studied into two homogeneous groups. The two regimes T3 (100% ETc) and T2 (70% ETc) formed a single homogeneous group, while regime T3 (PA) formed the other group. Consequently, the effect of these two irrigation regimes on stomatal conductance was identical. The low values of stomatal conductance obtained in the T3 (PA) treatment could be explained by the fact that the young trees subjected to this irrigation regime are exposed to a water deficit compared with the other two treatments. Similar results were reported by (Fernández et al., 1997) for the Manzanille variety under deficit irrigation. Stomata generally close when a plant is subjected to water stress, in order to reduce transpiration and improve the water status of its tissues.

II.4. Effect on leaf soluble sugar content

Figure 10 shows that the soluble sugar content varies according to the water regime. These differences were highly significant between the three treatments studied (P<0.001 at 2 dll) (table 6 appendix).

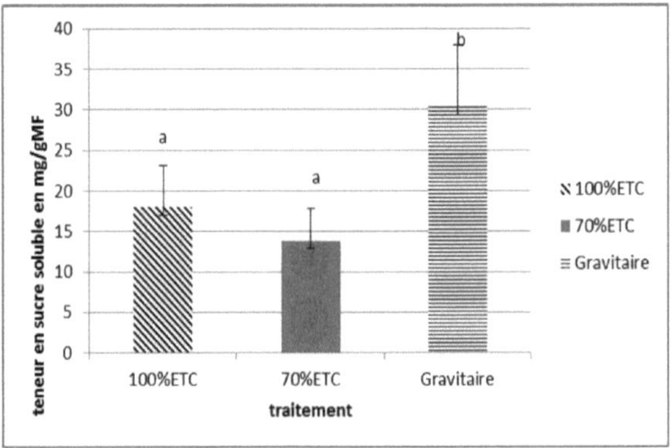

**Figure 10: Effect of water regime on soluble sugar content in leaves of
young olive trees**

The lowest levels of soluble sugars were observed in the young trees subjected to treatment T2 (70% ETc), while the highest levels were found in the trees subjected to treatment T3 (PA). The SNK test (Appendix Table 15) classified the treatments studied into two homogeneous groups. The first group includes treatments T3 (100%ETc) and T2 (70%ETc), and the second group includes treatment T3 (PA). The high accumulation of soluble sugars observed in young trees in the latter treatment could be an indicator of water stress induced by the gravity irrigation system used by farmers. Indeed, sugars are among the important osmolytes that contribute to the osmotic adjustment of plants during a water or salt deficit (Ashraf and Harris, 2004).

II.5. Effect on proline content in leaves

Analysis of variance showed a highly significant effect of the irrigation treatments on proline accumulation in the leaves (P<0.001 2 ddl)(table7 appendix). Examination of figure 11 showed that the T3 (PA) irrigation regime caused proline accumulation in the leaves.

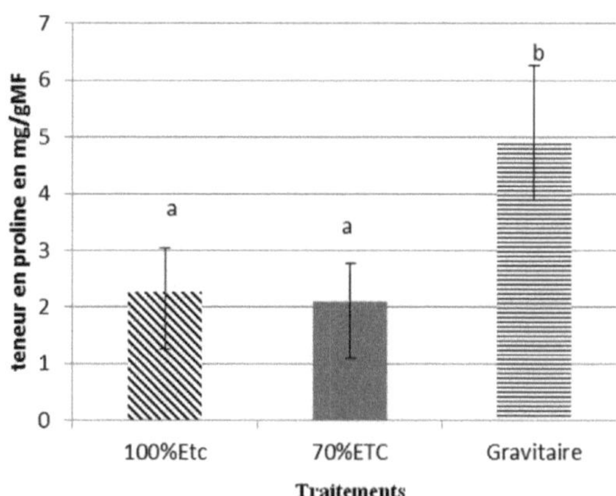

Figure 11: Effect of water regime on proline accumulation in the leaves of young olive trees

The SNK test (Appendix Table 16) classified the treatments studied into two homogeneous groups. The fact that the two regimes T1 (100% ETc) and T2 (70% ETc) belong to the same homogeneous group shows that the behaviour of these two regimes with respect to this hydric constraint was identical. Consequently, saving irrigation water in treatment T2 (70% ETc) did not induce water stress in the young olive trees. The accumulation of proline is known to be one of the most remarkable mechanisms of water stress. This amino acid acts as an osmoticum whose cytoplasmic accumulation lowers the water potential, thereby increasing the gradient for water intake and maintaining turgidity (Monneveux and This, 1997), which has also been observed in olive trees (Chartzoulakis et al., 1999; Dichio et al., 2005).

II.6. Effect on leaf protein content

The results shown in Figure 12 show that the effect of the water deficit caused a highly significant variation between the irrigation treatments in terms of leaf protein content (Appendix Table 9).

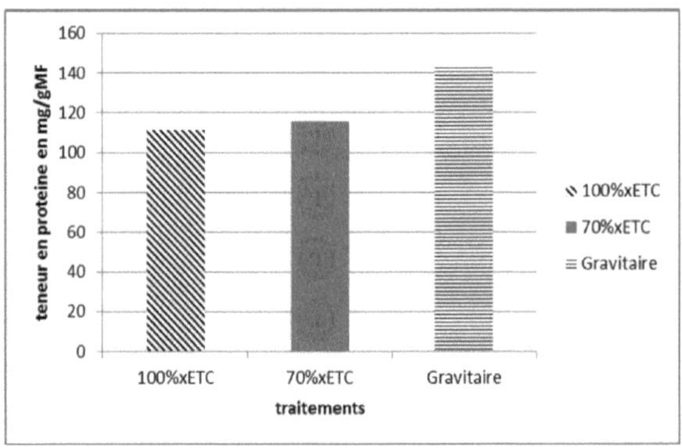

Figure 12: Effect of water regime on the protein content of young olive tree leaves

The SNK test (table 18 in the appendix) classified the treatments studied into two homogeneous groups. The two regimes T1 (100%) and T2 (70% ETc) belong to the same homogeneous group. This shows that there are no significant differences between these treatments. During a water deficit, the protein content of the olive tree increases, and several proteins may be indicative of a lack of water, such as dehydrin (Galau et close, 1992).

II.7. Effect on chlorophyll content in leaves

Figure 13 shows the effect of the water regime on chlorophyll accumulation in olive leaves. This accumulation of chlorophyll shows significant variations according to the irrigation treatments. Analysis of variance showed a high level of significant difference between treatments (Appendix Table 8). Treatment T3 (PA) caused a decrease in chlorophyll pigments in the leaves. The highest levels were found in trees subjected to Treatment T1 (100% ETc).

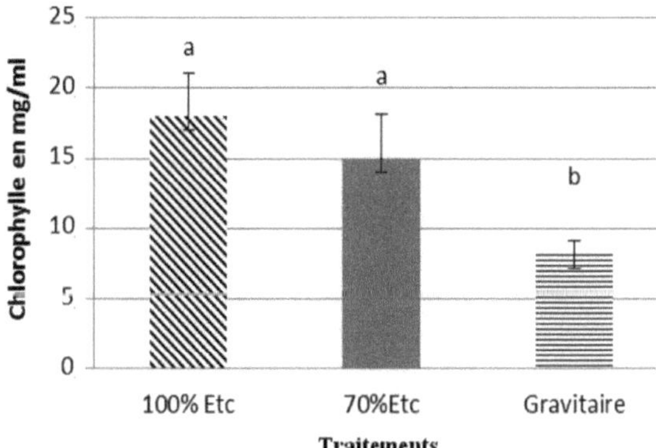

Figure 13: Effect of water regime on the chlorophyll content of the leaves of young olive trees.

Two homogeneous groups were identified by the SNK test (Table 17, Appendix). The two treatments T1 (100% ETc) and T3 (70% ETc) make up the same group, which shows that these two irrigation regimes behave identically in relation to this hydric constraint. The decrease in chlorophyll content noted in treatment T3 (PA) indicates that this irrigation regime exposes young olive trees to water stress. According to Pham Thi et al (1985), water deficit leads to a reduction in the concentration of chlorophyll pigments in the plant. This phenomenon can be interpreted either as the degradation of pigments by hydrolytic enzymes, or as the inactivation of the biosynthesis of these pigments.

CONCLUSION

The study of the effect of the irrigation regime on the agro-physiological parameters of young olive trees (Menara variety) was conducted at the INRA Saada Experimental Station. For the agronomic parameters, the T3 irrigation regime induced depressive effects on all the vigour parameters studied compared with the T3 treatments, which induced the highest values for these parameters. However, the T2 deficit irrigation (70%ETc) did not induce significant reductions in these parameters compared with the T1 treatment. As for the physiological parameters, the T3 irrigation regime induced a decrease in relative water content, stomatal conductance and chlorophyll, and an increase in the soluble sugar, protein and proline content of the leaves. This compared with the normal T1 diet (100%ETc). On the other hand, the results of the physiological parameters obtained by the T2 deficit irrigation (70% ETc) were not significantly different from those recorded by the T1 irrigation regime (100% ETc). Consequently, gravity-fed irrigation, as practised by farmers, does not meet the need to save water, which is becoming increasingly scarce as a result of climate change, nor does it ensure efficient use of young olive trees. On the other hand, T2 deficit irrigation (70%ETc) saves water without adversely affecting the agro-physiological parameters of young olive trees. Consequently, this irrigation regime should be recommended to replace the gravity irrigation regime in arid zones.

OUTLOOK

As with any scientific work, the results obtained open the way to several perspectives. The lines of research proposed to complete the study of irrigation deficiency in olive trees are as follows:

► Study the impact of deficient irrigation on lipogenesis and nutritional quality and organoleptic qualities of olive oil.

► Carry out a detailed mineral analysis to better understand the effect of water deficit on trophic relationships in olive trees.

► Determination of other metabolites that may be involved in the processes of osmoregulation, namely soluble sugars and glycine betaine.

► Evaluation of the effect of this stress on proteins and enzyme activities involved in the various metabolic processes.

BIBLIOGRAPHICAL REFERENCES

Agabbio, 1974 ,influenza dell'invtervento irriguo sul ciclo productivo dell'olivico pp 300-398

Arnon, D.I., 1949. Copper enzymes in isolated chloroplasts: ployphenol-oxidase in Beta vulgaris L., Plant Physiol. 24:1-15

Ashraf, M ., Harris, P.J.C., 2004. Potential biochemical indicators of salinity tolerance in plants. Plant. Sci. 166: 3-16.

Ashraf, M. Harris, P.J.C. (2004). Potential biochemical indicators of salinity tolerance in plants. Plant Science, 166, 3-16

Azouggagh M., 2001. Transfer of technologie enagriculture : bulletin monthly newsletter MADREEF /DERD N 81.

Barranco, D., Rallo, L., 1995.Las variedades de oliveo coltivadas en Andalucia.Juntade Andalucia, conserjeri de agricultura,pesca y alimentacion.Cordoba,Spain.

Bekkar, Y., Kuper, M., Hammani, A., Dionnet, M., Eliamani, A., 2007. "Reconversion to localised irrigation systems in Morocco. Quel enseignements pour l'agriculture familiale?" Revue HTE N°137, p. 38.

Bhatt, R.M., Srinivasa, Rao, N.K., (2005). Influence of pod load response of okra to water stress. Indian J. Plant Physiol, 10: 54- 59.

Binzel, M.L., Bressan, R.A., Handa, S., Handa, A.K., Hasegawa, P.M., Rhodes, D., 1987. Solute accumulation in tobacco cells adapted to NaCl. Plant physiology. 84: 1408-1415.

Bouat, A., 1980. L'analyse végétale dans le contrôle de l'alimentation des plantes tempérés et tropical. Ed. Lavoisier, Paris, 810p.

**Boulouha, B., Sikaoui, L., Hadiddou, A., Mamouni, A., Ougas, Y., Oukabli,
2006."Technical info: Olive trees. Installation and management of
cultivation. Published by INRA, pp. 5-14.**

Boulouha, 1986.fruiting growth and their interaction on production i n
Moroccan picholine. Oleap, pp. 41-46.

Boulouha, 2006 International seminar on the olive tree. Genetic improvement
Bradford, M.M., 1976. A rapid and sensitive method for the quantification of
microgram qauntities of protein utilizing The principal of protein dye binding.
Anal. Bioch. 72: 248-257.

C.O.I (International Olive Oil Council), 2007. "Techniques de production en
oléiculture". pp, 23, 146.

Camps, G., 1974. Les civilisations préhistoriques d'Afrique du nord et du
Sahara. Paris, Doin,. pp. 51 and 90.

Chartzoulakis, K., Patakis, A., Bosabalidis, A.M., 1999. Changes in water
relations, photosynthesis and leaf anatomy induced by intermittent drought in
two olive cultivars. Environ. Exp. Bot. 42: 113-120.

Christiansen, M.M., 1982. World environmental limitations to food and fibre
culture. Wilded Interscience, pp. 1-11.

Dichio, B., Xiloyannis, C., Sofo, A., Montanaro, G., 2005. Osmotic regulation
in leaves and roots of olive trees during a water deficit and rewatering. Tree
Physiology 26, 179-185.

Dubois, F., Gilles, K.A, Hamilton, J.K., Rebers, P.A. Smith, F., 1956.
Colorimetric method for determination of sugars and related substances. Anal.

Chem. 28: 350-356.

Ennajah, M., Vadel, A.M., Khemira, H., BenMimoun, M., Hellali, R., 2006. Defense mechanisms against water deficit in two olive (Olea europeae L.) cultivar 'Meskia and 'Chemlali'. J. Hotic. Sci. Biotech. 81 : 99-104.

Farissi M., Bouizgaren A., Faghire M., Bargaz A. & Ghoulam C. 2013. Agro- physiological and biochemical properties associated with tolerance of Medicago sativa populations to water deficit, **Turkish Journal of Botany**, 37, 1166-1175.

Farooq, M., Wahid, A., Kobayashi, N., Fujita, D., Basra, S.M.A., (2009). Plant drought stress: effects, mechanisms and management. Agron. Sustain. Dev. 29: 185-212.

Farouk, I.A., 2011. "Morphological and molecular study of 14 genotypes obtained by cross-breeding between domestic and foreign olive varieties", p.5.

Fereres, E., 1984. Variability in adaptive mechanisms to water deficits in annual and perennial crop plants. Bulletin de la Société Botanique de France. Actualités Botaniques 131: 17 32.

Fereres, E., Pruitt, W.O., Beutel, J.A., Henderson, D.W., Holzapfel, E., Shulbach, H., Uriuk, K., 1981. ET and drip irrigation scheduling. In: Fereres, E. Ed. Drip irrigation management. University of California. Div. of Agric. Sci. No. 21259, pp. 8-13.

Fereres, E., Ruz, C., Castro, J., Gomez, J.A., Pastor, M., 1996. Recuperation del olivo despues de una sequia extrema. XIV Congreso Nacional de Riegos. Almeria.

Fereres, F., 1984. Variability in adaptive mechanisms to water deficits in annual and perennial crop plants. Bulletin Société Botanique de France. Actualités Botaniques, pp. 131: 17-32.

Fernández, J.E., Moreno, F., Girón, I.F. Blázquez, O.M., 1997. Stomatal control of water use in olive tree leaves. Plant and Soil, 190: 179-192

Fernandez, J.E., Morino, F., 1999. Water use by the olive tree. J. Crop. Prod. 2, 101-162.

Filali, B.A., 2010. "Drip irrigation systems: development, operation,

Galau, G., Close, T.J., 1992. Sequence of the cotton group 2 LEA/ RAB/ dehydrin proteins en encoded by lea3 cDNCs. Plant physiology. 98: 1523-1525.

Hoekstra, F.A., Golovina, E.A., Buitink, J., 2001. Mechanisms of plant desiccation tolerance. Trends. Plant. Sci. 9: 431-438.

Kotchi, S.O., 2004. Detection of water stress by infrared thermography: Application to potato cultivation.

Kusaka, M., Ohta, M., Fujimura, T., 2005. Contribution of inorganic components to osmotic adjustment and leaf folding for drought tolerance in pearl millet. Plant physiology. 125: 474-489.

Loussert, R., Brousse, G., 1978. L'olivier. Technique agricole et productions méditerranéene, Edition nationale d'agriculture 10-11Décenbre 1986, Al Awamia N°68, pp. 103-113 ; 231-252.

Loussert, R., and Ferrak, A., 2011. Secret de l'olivier. Edition PCM consultivy, pp :199.

MADRPM (Ministry of Agriculture, Rural Development and Maritime Fisheries), December 2008. "Plan National Oléicole, Bulletin de liaison du programme national de transfert de technologie en agriculture".

Makela, P., Munns, R., Colmer, T.D., Condon, A.G., Peltonen - Sainio P., 1998. Effect of foliar applications of glycinebetaine on stoamatal conducatnce, abscissic acid and solute concentrations in leaves of salt or drought-stressed tomato. Aust. J. Plant physiol. 25: 655- 663.

MAPM (Ministry of Agriculture and Maritime Fishing). 2013. Filière oléicole (Official website of the ministry).

MAPM (Ministry of Agriculture and Maritime Fishing), 2013. Irrigation in Morocco (official Ministry website).

Mohamed Lamine, 1993 Agronomic diagnosis of olive tree management and maintenance in the Amizmiz region.

Mckersie, B.D., Leshan, Y.Y., 1994. Stress and stress coping in cultivated plants. Kluwer academic publishers, London.

Mehanna, H.T., Stino, R.G., Ikram, S.E., Gad El-Hak, A.H., (2012). The Influence of Deficit Irrigation on Growth and Productivity of Manzanillo Olive Cultivar in Desert Land. Journal of Horticultural Science & Ornamental Plants 4 (2): 115-124, 2012

Michelakis, N., 1998. Water management and irrigation for olive trees. Proceedings of the international seminar on olive growing, held 1997, Greece.

Mohammadian, R., Mghaddam, M.R., Sadeghian, S.Y., 2005. Effect of early season drought stress on growth characteristics of sugar beet genotypes. 29: 357- 368.

Monneveux, P., This, D., 1997. La génétique aux problèmes de la tolérance des plantes cultivées à la sécheresse: espoirs et difficultés. 8: 29-37.

Nemmar, M., 1983. Contribution à l'étude de la résistance à la sècheresse chez les variétés de blé dur (Triticum durum. Desf) et de blé tendre (Triticum aestirum L.) Evolution des teneurs en proline au cours du cycle de développement", thesis Doct Ing. onpellier, P.108

Nieves, N., Martinez, M.E., Castillo, R., Blanco, M.A., Gonzàlez-olmedo, J.L., 2001. Effect of abscisic acid and jasmonic acid on partial desiccation of encapsulated somatic embryos of sugarcane. Plant cell tiss. Org. cult. 65: 15-21.

Nonami, H., (1998). Plant water relations and control of cell elongation at low water potentials.J. Plant Res, 111: 373-382.

Orgaz, F., Fereres, E., 2007. Riego. In Barranco,D., Fernandez Escobar, R. and Rallo, L. El cultivo del olivo. Co Edit. junta Andalucia et MP, 724, pp: 287-306.

ORMVAH (Office Régionale de Mise en Valeur Agricole du Haouz), 2005. Guide de **l'irrigation dans le Haouz et fiche régionale de mise en valeur agricole du Haouz.**

Paquin R., Lechasseur P., 1979 - Observations sur une méthode de dosage de la praline libre dans les extraits de plantes, Can J Bot 57: 1851-1854.

Pham Thi, A.T., Borrel-Floud, C., Vieira, S.J., Justin, A.M., Mazliak, P., 1985. Effects of water stress on lipid metabolism in cotton leaves. Phytochemistry. 24: 723-727.

Riccardi, F., Gazeau, P., De Vienne, D., Zivi, M., 1998. Protein changes in response to progressive water deficit in maize: quantitative variation and polupeptide identification. Plant physiology. 117: 125-126.

Seyed, Y.S.L., Rouhollah, M., Mosharraf, M.H., Ismail, M.M.R., 2012. Water Stress in Plants: Causes, Effects and Responses, Water Stress, Prof. Ismail Md. Mofizur Rahman (Ed.), ISBN: 978-953-307-963-9, InTech, Available from: http://www.intechopen.com/books/water- stress/water-stress-inplants-causes-effects-and-responses

Trossat, C., 2005. Wheat water stress course. France.

William, G.H., Charles-Marie, E., 2003. Plant physiology, p. 456.

Farissi M., Bouizgaren A., Faghire M., Bargaz A., Ghoulam C., 2011. Agro-physiological responses of Moroccan alfalfa (Medicago sativa L.) populations to salt stress during germination and early seedling stages. Seed Science and Technology. 39: 389-401.

Coudret A. 1981. Action of NaCI on water stress and relations in the aerial parts of Plantago maritima L. and Plantago lanceolata L. Oecol. Plantarum, 2, 16, 111-120.

APPENDIX

Table 1: ANOVA table for average tree height

Parameter	SCE	Ddl	CM	Fisher	Meaning
Total height	32034,754	2	16017,377	19,223	<0,001***

***** Highly significant (P<0.001); * Significant (P<0.05)**

Table 2: ANOVA table for average tree diameter

Parameter	SCE	Ddl	CM	Fisher	Meaning
Average diameter	36808,711	2	18404,355	7,238	<0,001***

Table 3: ANOVA table for trunk section area

Parameter	SCE	Ddl	CM	Fisher	Meaning
Trunk perimeter	300,685	2	150,343	9,593	<0,001***

***** Highly significant (P<0.001); * Significant (P<0.05)**

Table 4: ANOVA table for stomatal conductance

Parameter	SCE	Ddl	CM	Fisher	Meaning
Stomatal conductance	37614,318	2	18807,159	29,931	<0,001***

***** Highly significant (P<0.001); * Significant (P<0.05)**

Table 6: ANOVA table for soluble sugar content

Parameter	SCE	ddl	CM	Fisher	Meaning
Soluble sugar content	828,453	2	414,227	13,476	<0,001***

***** Highly significant (P<0.001); * Significant (P<0.05)**

Table 7: ANOVA table for proline content

Parameter	SCE	ddl	CM	Fisher	Meaning
Proline teat	27,556	2	13,778	16,125	<0,001***

Table 8: ANOVA table for chlorophyll content

Parameter	SCE	ddl	CM	Fisher	Meaning
Chloropylle content	324,985	2	162,492	21,922	< 0,001***

Table 9: ANOVA table for protein content

Parameter	SCE	ddl	CM	Fisher	Meaning
Protein content	4249,537		2124,768	3,647	P (a=5%). *: significant (P<0.05)

Table10 SNK test: height

TREES	N	SUBSET	
		1	2
Grav	40	220,10	
70%ETc	63		248,43
100%Etc	63		255,24
Sig.		1,000	,227

Table 12 SNK test: trunk perimeter

trees	N	Subset	
		1	2
Grav	40	17,063	
70%ETc	63	19,913	
100%Etc	63	20.421	
SIG	1.000	510	

Table11 SNK test: average diameters

trees	N	Subset	
		1	2
Grav	40	141,813	
70%ETc	63	174.159	
100%Etc	63	100%Etc	
SIG	1.000	656	

Table13 SNK test: Conductance mmol/(m²-s)

Treatment	N	Subset	
		1	2
Gravity	17	74,641	
100%ETC	22		129,850
70%ETC	22		130,200
Sig.		1,000	,965

Table 15 SNK test: Sugar content

treatment	N	Subset	
		1	2
70%ETC	7	13,80100250626570	
100%ETC	8	17,97824561403510	
Gravity	5		30,29592982456140
Sig.		,195	1,000

Table 16 SNK test: Proline content			
Treatments	N	Subset	
		1	2
70%ETC	8	2,10008333333333	
100%Etc	7	2,27081904761905	
Gravity	5		4,88517333333333
Sig.		,745	1,000

Table 17 SNK test: Chlorophyll content

trt	N			Subset		
			1		2	
Gravity		6		8,17		
70%Etc		9			15,00	
100% Etc		7			18,00	
Sig.				1,000	,051	

Table18 SNK test: Protein content

Treatment	N		Subset
		1	2
100%xETC	9	111,08392796466200	
70%xETC	9	115,33809038396200	
Gravity	6		143,64525993883800
Sig.		,733	1,000

Printed by Books on Demand GmbH, Norderstedt / Germany